The Future of Learning Educational Data Mining for Every Individual

Edwin Kane

Copyright © [2023]

Title: The Future of Learning Educational Data Mining for Every Individual
Author's: Edwin Kane

This book was printed and published by [Publisher's: **Edwin Kane**] in [2023]

ISBN:

TABLE OF CONTENT

Chapter 1: Introduction to Educational Data Mining

Understanding Educational Data Mining

Importance of Educational Data Mining in Learning

Benefits and Challenges of Educational Data Mining

Ethical Considerations in Educational Data Mining

Chapter 2: Foundations of Educational Data Mining

Introduction to Data Mining Techniques

Data Preprocessing and Cleaning

Data Visualization and Exploration

Statistical Analysis in Educational Data Mining

Chapter 1: Introduction to Educational Data Mining

Understanding Educational Data Mining

In the rapidly evolving world of education, the extraction of valuable insights from data has become more crucial than ever before. Educational Data Mining (EDM) is a powerful tool that enables educators, researchers, and policymakers to uncover patterns, trends, and relationships hidden within educational data. By leveraging these insights, stakeholders can make informed decisions to enhance learning outcomes and improve the overall educational experience for every individual.

EDM utilizes data mining techniques specifically tailored to the field of education. It involves collecting, cleaning, analyzing, and interpreting large volumes of data generated within educational settings. This data could include information from student assessments, attendance records, course materials, and even data from online learning platforms.

The primary goal of EDM is to transform raw data into actionable knowledge. By employing various statistical and machine learning algorithms, educational researchers can identify patterns that may not be easily discernible through traditional methods. These patterns can then be used to develop personalized learning strategies, identify at-risk students, and evaluate the effectiveness of educational interventions.

For educators, EDM offers valuable insights into student behavior and performance. By analyzing data, teachers can gain a deeper understanding of how students learn, their strengths, and areas where additional support may be required. This knowledge allows educators to tailor their instruction to meet individual student needs, making the learning process more engaging and effective.

Researchers also benefit from EDM by being able to conduct large-scale studies to investigate educational phenomena. By analyzing data from diverse populations, researchers can identify factors that contribute to student success and develop evidence-based recommendations for educational policies and practices.

Policymakers can utilize EDM to make data-driven decisions that impact educational systems at a broader level. By analyzing aggregated data from multiple schools or districts, policymakers can identify systemic issues, develop targeted interventions, and allocate resources more effectively.

In conclusion, Educational Data Mining is a powerful tool that has the potential to revolutionize education. By understanding and harnessing the insights hidden within educational data, we can empower educators, researchers, and policymakers to make informed decisions that benefit every individual. The future of learning lies in the effective use of educational data mining, and its impact will continue to shape the educational landscape for years to come.

Importance of Educational Data Mining in Learning

Subchapter: Importance of Educational Data Mining in Learning

Introduction:

In the digital age, data mining has become an indispensable tool for various industries, and education is no exception. Educational Data Mining (EDM) is a field that utilizes data mining techniques to extract valuable insights from educational data. By analyzing vast amounts of information collected through educational systems, EDM helps educators and learners make informed decisions, tailor instruction, and enhance the learning experience. This subchapter explores the significance of EDM in learning and its potential to revolutionize education for everyone.

Understanding Learners:

EDM enables educators to gain a deep understanding of learners' behaviors, preferences, and progress. By analyzing data from various sources, such as learning management systems, online platforms, and assessments, educators can identify patterns and trends in students' learning habits. This knowledge enables educators to personalize instruction, identify struggling students, and develop targeted interventions. By understanding learners on an individual level, EDM empowers educators to provide tailored support and enhance learning outcomes.

Predictive Analytics:

One of the key benefits of EDM is its ability to make accurate predictions about learner performance. By analyzing historical data, EDM algorithms can identify factors that contribute to success or failure. These predictive analytics help educators identify at-risk students early on and implement proactive measures to support them. By leveraging these insights, educators can intervene before a student falls behind, ensuring a more inclusive and equitable learning environment for all.

Curriculum Development:

EDM also plays a significant role in curriculum development. By analyzing data on learner performance and engagement, educators can identify gaps in the curriculum and areas where instructional materials may need improvement. This data-driven approach to curriculum development ensures that educational resources are aligned with learners' needs and helps educators create more effective and engaging learning experiences.

Continuous Improvement:

EDM promotes a culture of continuous improvement in education. By collecting and analyzing data on instructional strategies, assessments, and learner outcomes, educators can identify effective teaching practices and refine their instructional methods. This iterative process allows educators to adapt and evolve their teaching strategies based on real-time feedback, resulting in improved learning outcomes for all students.

Conclusion:

Educational Data Mining is rapidly transforming education by harnessing the power of data to enhance learning experiences. By understanding learners, predicting performance, informing curriculum development, and fostering continuous improvement, EDM has the potential to revolutionize education for every individual. As we embrace the future of learning, EDM will continue to play a crucial role in creating personalized, inclusive, and effective educational experiences for learners of all ages and backgrounds.

Benefits and Challenges of Educational Data Mining

In today's rapidly evolving digital era, educational institutions are generating an unprecedented amount of data. This valuable resource has the potential to revolutionize the way we learn and teach. Educational Data Mining (EDM) is the process of extracting meaningful patterns and insights from this data, offering numerous benefits for students, teachers, and educational organizations. However, along with these benefits come certain challenges that need to be addressed.

One of the key benefits of EDM is its ability to personalize learning experiences. By analyzing vast amounts of data on individual students, EDM can identify their strengths, weaknesses, and learning preferences. This enables educators to tailor their teaching methods accordingly, ensuring that each student receives a personalized and effective education. No longer will students be limited by a one-size-fits-all approach, as EDM allows for adaptive learning paths that accommodate their unique needs.

Another advantage of EDM is its potential to improve student outcomes. By analyzing data on student performance, educators can identify patterns that lead to academic success. This information can then be used to develop targeted interventions and strategies to help struggling students. Additionally, EDM can provide early warning systems that alert educators to students who may be at risk of dropping out or underperforming, allowing for timely intervention and support.

Moreover, EDM can enhance the overall efficiency of educational institutions. By analyzing data on resource allocation, course scheduling, and student enrollment trends, administrators can make informed decisions that optimize the use of resources and improve institutional performance. This data-driven approach can lead to cost savings, improved student satisfaction, and better overall outcomes for educational organizations.

However, despite its numerous benefits, EDM also presents certain challenges. One of the primary concerns is the ethical use of student data. Safeguarding student privacy and ensuring data security are paramount in the implementation of EDM. Striking a balance between data collection and privacy protection is crucial to maintain trust between educational institutions, students, and their families.

Another challenge lies in the interpretation and utilization of data. While EDM can provide valuable insights, it is important for educators and administrators to have the necessary skills and knowledge to effectively analyze and interpret the data. Training and professional development programs should be implemented to equip educators with the skills needed to make informed decisions based on the data generated by EDM.

In conclusion, Educational Data Mining holds tremendous potential to transform the future of learning. Its ability to personalize education, improve student outcomes, and enhance institutional efficiency makes it an invaluable tool for educational organizations. However, challenges such as data privacy and interpretation must be carefully addressed to ensure its effective implementation. By harnessing the power of EDM, we can unlock a new era of education that caters to the unique needs of every individual learner.

Ethical Considerations in Educational Data Mining

In the rapidly advancing field of educational data mining, where sophisticated algorithms and data analytics are being used to analyze vast amounts of educational data, it is crucial to address the ethical considerations that arise. Educational data mining (EDM) holds great potential for improving learning outcomes and personalizing education for every individual. However, as with any powerful technology, there are ethical dilemmas that need to be carefully navigated to ensure the responsible and ethical use of data mining in education.

One of the key ethical considerations in educational data mining is privacy. As educational institutions collect and analyze large amounts of student data, there is a need to safeguard this information and ensure that it is used solely for educational purposes. It is imperative to obtain informed consent from students or their parents/guardians and to establish robust data protection measures to prevent unauthorized access or misuse of sensitive information. Additionally, data anonymization and aggregation techniques should be employed to protect individual identities while still enabling meaningful analysis.

Another ethical consideration is the potential for bias in educational data mining. Data mining algorithms are only as unbiased as the data they are trained on. If the data used for analysis is biased or reflects existing societal inequalities, the algorithms may inadvertently perpetuate these biases, leading to unfair treatment or discrimination. It is essential to constantly evaluate and address biases in the data, and to design algorithms that are fair and equitable for all individuals, regardless of their background or characteristics.

Transparency and accountability are also crucial ethical considerations in educational data mining. Educational institutions, researchers, and data scientists must be transparent about the data they collect, how it is used, and the algorithms employed. Openness and transparency build trust and allow for scrutiny and accountability in the use of educational data mining techniques.

Lastly, educational data mining should always prioritize the well-being and best interests of the learners. While data mining can provide valuable insights and personalized learning experiences, it is important to strike a balance between data-driven decision-making and the human touch in education. The ethical use of data mining should enhance the learning experience, empower learners, and promote their overall well-being.

In conclusion, ethical considerations are paramount in the field of educational data mining. Privacy, bias, transparency, and learner well-being are all critical aspects that must be carefully addressed to ensure the responsible and ethical use of data mining in education. By incorporating these considerations into the development and implementation of data mining techniques, we can harness the full potential of educational data mining while upholding ethical standards and benefiting every individual in the learning process.

Chapter 2: Foundations of Educational Data Mining

Introduction to Data Mining Techniques

Data mining is a powerful tool that has revolutionized the way we analyze and interpret large amounts of data. In today's digital age, where vast amounts of information are generated every second, it has become essential to extract valuable insights and patterns from this data to make informed decisions. This subchapter titled "Introduction to Data Mining Techniques" aims to provide a comprehensive overview of the fundamentals of data mining and its relevance in various industries.

Data mining refers to the process of discovering patterns, correlations, and trends within large datasets. It involves applying various statistical and mathematical techniques to extract meaningful information that can be used for predictive analysis, decision-making, and improving business strategies. The application of data mining techniques has been proven valuable in diverse domains such as finance, healthcare, marketing, and education.

In this subchapter, we will delve into the key data mining techniques that are widely used today. We will start by introducing the concept of data preprocessing, which involves cleaning, transforming, and integrating raw data to ensure its quality and applicability. Next, we will explore different types of data mining algorithms, including classification, clustering, association rule mining, and regression analysis. Each algorithm will be explained in a clear and concise manner, emphasizing its purpose and potential applications.

Furthermore, we will discuss the challenges and ethical considerations associated with data mining. Privacy concerns, data protection, and algorithm bias are some of the critical issues that need to be addressed when implementing data mining techniques. It is crucial for data scientists and analysts to be aware of these challenges and adopt responsible practices to ensure the ethical use of data mining technologies.

Finally, we will highlight the importance of data mining in the education sector. Educational data mining (EDM) is a specialized field that focuses on extracting valuable insights from educational data to enhance learning experiences and personalize instruction. We will explore how data mining techniques can be utilized to improve student performance, identify at-risk students, and optimize educational resources.

Whether you are a business professional, educator, or simply interested in the field of data mining, this subchapter will provide you with a solid foundation in understanding the basics of data mining techniques. By the end of this subchapter, you will have a clear understanding of the principles behind data mining and the potential it holds for transforming various industries, including education.

Data Preprocessing and Cleaning

In the rapidly evolving field of data mining, one crucial step that cannot be overlooked is data preprocessing and cleaning. This subchapter aims to provide a comprehensive understanding of this crucial process and its significance in educational data mining. Whether you are a seasoned data scientist or a beginner in the field of data mining, this subchapter is designed to cater to the needs of everyone interested in harnessing the power of data.

Data preprocessing involves transforming raw data into a format that is suitable for analysis. It is a critical step because the quality of the output is only as good as the quality of the input. Educational data mining relies on accurate and reliable data to derive meaningful insights and make informed decisions. Therefore, data cleaning becomes essential to ensure the integrity and reliability of the analysis.

This subchapter will delve into various techniques and methods used in data preprocessing and cleaning. It will cover important topics such as data integration, data transformation, data reduction, and data discretization. Each of these techniques plays a crucial role in preparing the data for analysis. The subchapter will also highlight common challenges faced during this process and provide practical solutions to overcome them.

Furthermore, this subchapter will explore the importance of data quality and the impact it has on the accuracy of the results. It will discuss data anomalies, missing values, and outliers – common issues faced when dealing with real-world datasets. Through practical examples and case studies, readers will gain a deeper understanding of how to identify and handle such challenges effectively.

Additionally, this subchapter will shed light on the ethical considerations of data preprocessing and cleaning in the field of educational data mining. It will emphasize the significance of data privacy, confidentiality, and informed consent when working with sensitive educational data.

Whether you are a data mining enthusiast, an educator, or a researcher, understanding the intricacies of data preprocessing and cleaning is essential. This subchapter aims to equip every individual interested in data mining with the necessary knowledge and skills to ensure the accuracy, reliability, and ethical usage of educational data. By mastering the art of data preprocessing and cleaning, you will be able to unlock the true potential of educational data mining and contribute to the future of learning.

Data Visualization and Exploration

In this rapidly evolving digital age, data has become an indispensable tool for understanding and improving educational systems. With the advent of educational data mining, we now have the ability to harness the power of data to gain valuable insights into individual learning processes. One of the key components of this process is data visualization and exploration.

Data visualization is a powerful technique that allows us to present data in a visually engaging and easily understandable format. By representing complex data sets through charts, graphs, and interactive visualizations, we can effectively communicate patterns, trends, and relationships that would otherwise be difficult to comprehend. This enables educators, researchers, and policymakers to make informed decisions and take necessary actions to enhance the learning experience for every individual.

Through data exploration, we dive deeper into the data to uncover hidden patterns and gain a comprehensive understanding of the learning environment. By using data mining techniques, we can extract valuable information from vast amounts of data collected from various sources such as learning management systems, online forums, and assessment tools. This information can then be visualized and analyzed to discover trends, identify knowledge gaps, and identify students who may be at risk of falling behind.

Data visualization and exploration have immense potential to revolutionize the field of education. For educators, it offers a way to track student progress, identify areas where additional support may be needed, and create personalized learning experiences tailored to each student's unique needs. Researchers can use these techniques to develop new theories, test hypotheses, and contribute to the growing body of knowledge in the field of educational data mining.

Furthermore, data visualization and exploration can foster collaboration between different stakeholders, including teachers, administrators, and policymakers. By presenting data in a format that is easy to understand, it promotes meaningful discussions and enables data-driven decision-making. This can lead to more effective interventions, improved resource allocation, and ultimately, better educational outcomes for every individual.

In conclusion, data visualization and exploration are essential tools in the field of educational data mining. They enable us to make sense of complex data sets, uncover hidden patterns, and gain valuable insights into the learning process. By harnessing the power of data, we can transform education and ensure that every individual receives the support and resources they need to succeed. Whether you are an educator, researcher, or policymaker, understanding and utilizing data visualization and exploration techniques is crucial in shaping the future of learning.

Statistical Analysis in Educational Data Mining

In the rapidly evolving landscape of education, data mining has emerged as a powerful tool to unlock the potential of educational data. By leveraging statistical analysis techniques, educators and researchers can gain valuable insights into student learning patterns, instructional effectiveness, and learning outcomes. In this subchapter, we explore the role of statistical analysis in educational data mining, shedding light on its significance and benefits for educators, students, and researchers alike.

Statistical analysis is an integral part of educational data mining, serving as the backbone for uncovering patterns and trends within large datasets. By applying various statistical models, educators can identify factors that influence student performance, such as demographics, prior knowledge, and learning styles. This information can then be used to tailor instructional approaches and interventions to cater to individual students' needs, fostering personalized learning experiences.

Moreover, statistical analysis allows educators to assess the effectiveness of different teaching strategies and interventions. By analyzing data on student performance, engagement, and feedback, educators can identify the most impactful instructional methods, leading to evidence-based decision-making. This not only improves student outcomes but also helps optimize educational resources and interventions, ensuring maximum efficiency and effectiveness.

For researchers, statistical analysis in educational data mining provides a wealth of opportunities to investigate complex educational phenomena. By analyzing large-scale datasets, researchers can uncover hidden patterns and relationships that may not be apparent through traditional methods. This enables them to develop and validate theories, contributing to the advancement of educational research and practice.

While statistical analysis in educational data mining holds immense potential, it also poses certain challenges. Ensuring the quality and reliability of data is crucial for accurate analysis. Additionally, privacy and ethical considerations must be carefully addressed to protect the rights and interests of students and other stakeholders involved.

In conclusion, statistical analysis plays a vital role in educational data mining, offering valuable insights and opportunities for educators, students, and researchers. By leveraging statistical models and techniques, educators can personalize learning experiences, optimize instructional strategies, and improve student outcomes. Researchers, on the other hand, can uncover new knowledge and contribute to the advancement of the field. As data mining continues to revolutionize education, statistical analysis will remain a cornerstone in harnessing the power of educational data for the benefit of every individual.

Chapter 3: Data Sources and Collection in Educational Data Mining

Types of Educational Data Sources

In the rapidly evolving field of educational data mining, it is crucial to understand the various sources of educational data available to researchers, educators, and policymakers. These data sources provide valuable insights into the learning process and can help inform decision-making to improve educational outcomes for every individual. In this subchapter, we will explore the different types of educational data sources and their significance in the context of data mining.

1. Learning Management Systems (LMS): LMS platforms, such as Moodle or Blackboard, store a wealth of data on student interactions, including course materials accessed, time spent on tasks, and quiz scores. LMS data provides valuable insights into student engagement, learning progress, and performance.

2. Assessment Data: This includes standardized tests, formative assessments, and quizzes. Assessment data allows for the evaluation of students' knowledge, skills, and abilities. Data mining techniques can uncover patterns in assessment results, aiding in the identification of areas where students may be struggling or excelling.

3. Social Media and Online Forums: With the increasing integration of technology in education, students engage in online discussion forums and social media platforms related to their studies. These platforms provide a rich source of data for understanding student collaboration, knowledge sharing, and social learning.

4. Academic Records: Student academic records contain information such as grades, attendance, course enrollment, and degree progress. By analyzing academic records, data mining can identify patterns that may help predict student success or failure.

5. Sensor Data: Technological advancements have allowed for the collection of sensor data, such as eye-tracking or brainwave data, during learning activities. This data provides insights into students' cognitive processes, attention levels, and emotional states during different learning tasks.

6. Learning Analytics Dashboards: Learning analytics platforms consolidate data from various sources, providing a comprehensive view of students' learning journeys. These dashboards enable educators to track students' progress, identify areas of improvement, and personalize learning experiences.

7. Open Educational Resources (OER): OER platforms offer a vast collection of freely accessible educational materials. Analyzing usage patterns and interactions with OER can provide insights into students' preferences, interests, and gaps in their learning.

Understanding the types of educational data sources available is essential for harnessing the power of data mining in education. By leveraging these data sources, researchers, educators, and policymakers can gain valuable insights into the learning process, identify areas for improvement, and ultimately enhance educational outcomes for every individual.

Collection Methods for Educational Data

In order to pave the way for a more personalized and effective education system, educational data mining has emerged as a powerful tool. By utilizing data mining techniques, educators and researchers can uncover valuable insights from large volumes of educational data. However, the success of any data mining project heavily relies on the collection methods used to gather the necessary data. This subchapter explores the various collection methods for educational data, ensuring that the future of learning is driven by accurate and meaningful information.

One of the primary sources of educational data is student records. These records contain vital information such as grades, attendance, and demographic details. Schools and educational institutions have extensive databases that store student records, making it a rich source for data mining purposes. Furthermore, online learning platforms and educational apps have become increasingly popular, providing an additional avenue for data collection. These platforms can track students' progress, engagement levels, and even their learning preferences, providing a wealth of information for analysis and improvement.

Surveys and questionnaires are another valuable method for collecting educational data. By directly asking students, parents, and teachers about their experiences, preferences, and opinions, researchers can gain qualitative insights that complement quantitative data. Surveys can be conducted in person, through email, or even through online platforms, allowing for a wide reach and ease of data collection.

Educational experiments and studies also serve as a crucial source of data for educational data mining. By designing controlled experiments and interventions, researchers can collect data on the impact of specific teaching methods, technologies, or interventions. This data can then be analyzed to identify effective strategies and improve educational practices.

Lastly, data mining can also benefit from the use of sensors and tracking devices. These devices can capture various aspects of a learner's behavior, such as eye movements, brain activity, or physical movements. By collecting this data, researchers can gain deeper insights into the learning process, identifying patterns and individual differences that were previously inaccessible.

In conclusion, the collection methods for educational data are diverse and ever-evolving. By leveraging student records, surveys, experiments, and cutting-edge technologies, educational data mining can unlock valuable insights into the learning process. These insights can be used to personalize education, improve teaching methods, and ultimately enhance the educational experience for every individual. The future of learning is data-driven, and the collection methods discussed in this subchapter are essential for realizing this vision.

Privacy and Security in Educational Data Collection

In today's digital age, educational institutions are increasingly relying on data collection and analysis to improve learning outcomes for every individual. With the advent of educational data mining, educators can gain valuable insights into student performance, preferences, and areas of improvement. However, as we delve deeper into the realm of data mining, it is crucial to address the concerns surrounding privacy and security.

Privacy is a fundamental right that cannot be compromised, even in the pursuit of educational excellence. Educational data collection should be conducted within a framework that respects and safeguards the privacy of every individual. This means that any personally identifiable information should be handled with utmost care and stored securely. Educational institutions must adopt strict protocols to protect student data from unauthorized access, misuse, or breaches.

To ensure privacy, educational institutions should obtain informed consent from students and their parents or guardians before collecting any personally identifiable information. This consent should clearly outline the purpose of data collection, the types of data being collected, the duration of data retention, and the measures taken to protect the data. Additionally, institutions should provide individuals with the right to access and correct their data and should seek consent for any third-party sharing of data.

The security of educational data is equally important. Educational institutions must invest in robust cybersecurity measures to protect student data from cyber threats. This includes implementing firewalls, encryption techniques, secure data storage, and regular security audits. Furthermore, institutions should educate their staff and students about data security best practices, such as creating strong passwords, avoiding phishing attempts, and reporting any suspicious activities.

In the field of data mining, privacy-enhancing techniques can be employed to extract meaningful insights while preserving anonymity. Techniques like data anonymization, aggregation, and de-identification can be utilized to protect individual identities while still allowing for effective analysis. It is crucial for institutions to implement these techniques to strike a balance between data-driven decision-making and privacy preservation.

In conclusion, as the field of educational data mining continues to advance, it is essential to prioritize privacy and security. Educational institutions must implement stringent protocols and security measures to protect student data from unauthorized access or breaches. By respecting privacy rights and employing privacy-enhancing techniques, we can ensure that data mining in education remains a powerful tool for enhancing learning outcomes while maintaining the trust and confidence of every individual.

Data Quality and Reliability in Educational Data Mining

In the ever-evolving field of educational data mining, the quality and reliability of data play a crucial role in ensuring accurate analysis and meaningful outcomes. As technology continues to revolutionize the way we learn and teach, data mining has emerged as a powerful tool for extracting valuable insights from educational data. However, without reliable and high-quality data, the effectiveness and usefulness of educational data mining efforts can be severely compromised.

Data quality refers to the accuracy, completeness, consistency, and timeliness of the data being utilized. In the context of educational data mining, it is essential to ensure that the data collected is reliable and reflects the true characteristics of the learning process. This includes aspects such as student demographics, performance metrics, learning behaviors, and interactions with educational technology.

Reliability, on the other hand, focuses on the consistency and stability of the data over time. It is imperative to gather data consistently and continuously to identify patterns and trends accurately. In educational data mining, reliable data allows for the identification of long-term learning patterns, the evaluation of interventions, and the measurement of learning outcomes.

Ensuring data quality and reliability in educational data mining requires a multi-faceted approach. Firstly, data collection methods should be carefully designed to capture accurate and comprehensive information. This may involve the use of surveys, assessments, and learning analytics tools. The data collection process should also be standardized to ensure consistency across different educational settings.

Data cleaning and preprocessing techniques are equally important in maintaining data quality and reliability. This involves removing duplicate entries, handling missing data, and addressing any inconsistencies that may arise. Advanced statistical methods can be employed to validate and verify the accuracy of the data, ensuring that it represents the true characteristics of the learning environment.

Additionally, establishing data governance frameworks and protocols is essential to maintain data quality and reliability. This includes defining clear guidelines for data collection, storage, and usage, as well as ensuring compliance with privacy and ethical considerations. Adhering to these frameworks not only enhances data quality but also contributes to building trust among stakeholders in the educational data mining process.

In conclusion, data quality and reliability are critical pillars in educational data mining. By ensuring the accuracy, completeness, consistency, and timeliness of data, educational stakeholders can harness the full potential of data mining techniques to better understand and improve the learning experience. With careful attention to data collection, cleaning, and governance, educational data mining can pave the way for personalized and effective learning for every individual.

Chapter 4: Data Analysis Techniques in Educational Data Mining

Descriptive Analysis in Educational Data Mining

In the ever-evolving field of education, the role of data mining has become increasingly significant. Educational Data Mining (EDM) refers to the process of extracting valuable insights and patterns from educational data sets to inform decision-making and improve learning outcomes. One of the key techniques employed in EDM is descriptive analysis, which helps us gain a comprehensive understanding of the data.

Descriptive analysis in educational data mining involves the exploration and summarization of data to identify trends, patterns, and relationships. This technique provides educators, administrators, policymakers, and researchers with crucial information to design effective interventions, personalize instruction, and enhance educational systems.

For every individual, understanding descriptive analysis in educational data mining can be empowering. It allows us to make informed decisions about our own learning journey or that of our children. By examining large datasets, we can identify areas of strength and weakness, discovering which concepts are most challenging or where improvement is needed. With this knowledge, we can tailor learning strategies and resources to address specific needs, ultimately fostering a more personalized and effective educational experience.

Data mining enthusiasts and professionals will find descriptive analysis particularly intriguing. It provides them with the tools and techniques necessary to extract invaluable insights from educational data sets. From identifying common misconceptions to detecting learning preferences, descriptive analysis uncovers hidden patterns that can shape instructional design and curriculum development. By applying data mining principles, educators and researchers can optimize the learning process, leading to better student engagement and achievement.

Furthermore, descriptive analysis has the potential to revolutionize educational systems on a larger scale. By analyzing aggregated data from diverse populations, policymakers can gain a comprehensive understanding of educational trends and challenges. This information can inform evidence-based decisions to address systemic issues, allocate resources effectively, and implement targeted interventions. Through the integration of data mining techniques, educational institutions can work towards creating equitable and inclusive learning environments for all.

In conclusion, descriptive analysis plays a vital role in educational data mining. It enables individuals to make informed decisions about their learning journey, empowers educators and researchers to optimize instructional strategies, and supports policymakers in creating inclusive educational systems. By embracing the power of data mining, we can pave the way for a future of learning that is personalized, effective, and accessible to every individual.

Predictive Analysis in Educational Data Mining

In today's digital era, the field of education has witnessed a significant transformation with the advent of educational data mining (EDM). This innovative approach utilizes data mining techniques to extract valuable insights from vast amounts of educational data. One of the key applications within EDM is predictive analysis, which holds tremendous potential to revolutionize the way we understand and enhance the learning process.

Predictive analysis in educational data mining involves the use of statistical models and machine learning algorithms to forecast future outcomes and trends based on historical data. By analyzing patterns and associations in educational data, educators and policymakers can gain valuable insights into student behavior, learning patterns, and performance. These insights can then be used to predict student outcomes, identify at-risk students, and tailor interventions to improve educational outcomes.

The power of predictive analysis lies in its ability to identify patterns that may not be immediately visible to human observers. By analyzing a wide range of variables such as student demographics, previous academic performance, engagement levels, and learning styles, predictive analysis can provide valuable information about which students are likely to struggle or excel. This enables educators to intervene early and provide targeted support to those who need it most.

Furthermore, predictive analysis can also be used to personalize the learning experience for individual students. By leveraging data on students' strengths, weaknesses, and learning preferences, educators can tailor instructional strategies and resources to meet the unique needs of each student. This personalized approach not only enhances student engagement but also improves learning outcomes by ensuring that students receive the right support at the right time.

In addition to its benefits for individual students, predictive analysis in educational data mining also has the potential to inform policy decisions at a systemic level. By analyzing trends and patterns across large datasets, policymakers can identify areas of improvement, allocate resources effectively, and design evidence-based interventions to address educational challenges at scale.

Overall, predictive analysis in educational data mining offers immense potential to transform education by leveraging the power of data to improve learning outcomes for every individual. By harnessing the insights provided by predictive analysis, educators and policymakers can make informed decisions, personalize the learning experience, and ensure that every learner has the opportunity to reach their full potential. As data mining continues to evolve, the future of learning holds great promise for every stakeholder in the field of education.

Clustering and Classification Techniques

In the rapidly evolving field of educational data mining, clustering and classification techniques play a vital role in analyzing vast amounts of information to gain valuable insights into individual learning patterns. These techniques enable us to identify commonalities, group similar data points, and predict future behaviors, ultimately leading to more personalized and effective learning experiences for every individual.

Clustering techniques in data mining involve the grouping of data points based on similarities or shared characteristics. By clustering students with similar learning profiles together, educators can create tailored interventions and curriculum adjustments that address the specific needs of each group. This approach allows for more targeted instruction and a better understanding of the diverse learning styles within a classroom or educational system.

Classification techniques, on the other hand, are used to predict or classify new data based on patterns observed in existing data. By training algorithms on historical data, educators can create models that automatically categorize students into predefined groups or predict their future performance. These techniques enable early identification of struggling learners, allowing educators to intervene and provide additional support in a timely manner.

One of the main benefits of clustering and classification techniques in educational data mining is the ability to identify at-risk students and prevent learning gaps from widening. By analyzing various data sources such as student demographics, previous academic performance, and engagement patterns, educators can proactively identify students who may be falling behind or at risk of disengagement. With this information, targeted interventions can be implemented to address specific challenges and provide personalized support, ultimately improving student outcomes.

Furthermore, clustering and classification techniques can also facilitate personalized learning pathways. By understanding individual learning preferences, strengths, and weaknesses, educators can tailor instruction to meet the unique needs of each student. This approach fosters a more engaging and motivating learning environment, where students feel supported and empowered to take ownership of their learning journey.

In conclusion, clustering and classification techniques are powerful tools in educational data mining that enable us to uncover meaningful patterns and insights from large volumes of information. By utilizing these techniques, educators can gain a deeper understanding of individual learning profiles, identify at-risk students, and personalize instruction to maximize learning outcomes. As data mining continues to advance, these techniques will revolutionize the future of education, ensuring that every individual receives the personalized support and opportunities they deserve.

Association Rule Mining in Educational Data

Association rule mining is a powerful technique used in data mining to uncover relationships and patterns within datasets. In the context of education, association rule mining can provide valuable insights into student learning behaviors, instructional strategies, and curriculum design. This subchapter explores the application of association rule mining in educational data and its potential to revolutionize the future of learning.

Educational data mining, as a field, focuses on extracting knowledge from educational datasets to enhance teaching and learning processes. By analyzing large-scale educational data, such as student grades, assessments, and online interactions, educators and researchers can gain a deeper understanding of student behavior and make data-driven decisions to improve educational outcomes.

Association rule mining plays a crucial role in educational data mining by discovering interesting relationships and patterns in educational datasets. These rules can reveal hidden connections between various factors, such as student performance, engagement, demographics, and instructional methods. For example, association rule mining can uncover that students who frequently engage in online discussion forums have a higher likelihood of achieving better grades. Such insights can help educators identify effective teaching strategies and design personalized interventions for students at risk.

Furthermore, association rule mining can aid in course design and curriculum development. By analyzing student performance data, educators can identify patterns in successful learning sequences and identify the most effective instructional materials. This approach allows for a more personalized and adaptive learning experience, tailored to the individual needs and preferences of each student.

The application of association rule mining in educational data has the potential to transform traditional education systems. By leveraging the power of data mining techniques, educators can move away from a one-size-fits-all approach and embrace a more personalized and student-centered learning environment. Additionally, association rule mining can help identify early warning signs of student disengagement or learning difficulties, enabling timely interventions to support struggling students.

In conclusion, association rule mining is a valuable tool in the field of educational data mining. Its ability to uncover hidden patterns and relationships within educational datasets can revolutionize the future of learning. By leveraging association rule mining techniques, educators can make data-driven decisions, improve instructional strategies, and create personalized learning experiences for every individual. The integration of association rule mining in educational data has the potential to enhance educational outcomes and support the success of learners in various educational settings.

Chapter 5: Applications of Educational Data Mining

Personalized Learning and Adaptive Systems

In today's rapidly evolving educational landscape, the concept of personalized learning has gained significant attention. The traditional one-size-fits-all approach to education is being gradually replaced by a more tailored and individualized learning experience. This transformation has been made possible through the advancement of educational data mining and the development of adaptive systems.

Personalized learning refers to a student-centered approach that takes into account the unique needs, interests, and abilities of each learner. It recognizes that individuals have different learning styles and paces, and aims to provide them with customized learning experiences that meet their specific requirements. By leveraging educational data mining techniques, educators can gather valuable insights into students' learning patterns, preferences, and performance, enabling them to create personalized learning paths.

Data mining plays a crucial role in this process by analyzing large sets of educational data to discover meaningful patterns and relationships. It helps educators identify students' strengths and weaknesses, understand their learning behaviors, and predict their future learning trajectories. By utilizing data mining techniques, educators can make informed decisions about instructional strategies, content selection, and intervention methods. This empowers them to optimize the learning experience for each student, enhancing engagement, motivation, and ultimately, learning outcomes.

Adaptive systems, on the other hand, are technological tools that support personalized learning. These systems leverage the insights gained from data mining to dynamically adjust the learning experience based on the individual learner's needs. Adaptive systems can provide personalized content recommendations, adaptive assessments, and real-time feedback, all tailored to the learner's specific requirements. This level of customization ensures that learners are challenged appropriately and supported effectively, promoting continuous growth and improvement.

The integration of personalized learning and adaptive systems has the potential to revolutionize education. It provides learners with the opportunity to engage with educational materials in a manner that is most effective and enjoyable for them. It empowers educators to make data-driven decisions and deliver instruction that is responsive to each student's unique needs. Furthermore, it allows for greater flexibility and autonomy, enabling learners to take ownership of their educational journey.

In conclusion, personalized learning and adaptive systems are transforming education by tailoring the learning experience to meet the individual needs of learners. Through the application of data mining techniques and the utilization of adaptive systems, educators can create customized learning paths and provide personalized support to enhance learning outcomes. This shift towards personalized education has the potential to unlock the full potential of every learner and shape the future of learning for all.

Early Detection of Learning Difficulties

In the fast-paced world we live in, education plays a crucial role in shaping the future of individuals. However, not everyone finds the learning process easy and seamless. Some individuals face learning difficulties that hinder their progress and limit their potential. This is where the concept of early detection of learning difficulties comes into play, using the powerful tool of data mining.

Data mining is a technique that involves extracting valuable knowledge from large datasets. It has revolutionized various industries, and now, it holds immense potential in the field of education. By analyzing vast amounts of data collected from students, educators can identify patterns and trends that may indicate learning difficulties.

Early detection of learning difficulties is essential because it allows educators to intervene promptly and provide targeted support. By identifying struggling students early on, educators can design personalized learning plans to address their specific needs. This proactive approach helps prevent further frustration and disengagement, ultimately leading to improved learning outcomes.

Data mining techniques can be employed to identify various types of learning difficulties. For instance, it can help detect students who struggle with reading comprehension, mathematical concepts, or attention deficits. By analyzing data such as test scores, classroom behavior, and engagement levels, educators can gain insights into students' individual challenges.

Moreover, data mining can assist in identifying the underlying causes of learning difficulties. It can reveal patterns such as poor attendance, lack of engagement, or specific learning styles that may contribute to the struggles. Armed with this information, educators can implement targeted interventions and provide appropriate resources to help students overcome their difficulties.

However, it is important to emphasize that data mining in education is not meant to label students or stigmatize them based on their learning difficulties. Rather, it aims to empower educators with valuable insights to create inclusive and supportive learning environments.

In conclusion, the early detection of learning difficulties through data mining is a powerful tool that can revolutionize education. By analyzing vast amounts of data, educators can effectively identify struggling students, design personalized interventions, and ultimately enhance learning outcomes for every individual. With the future of learning relying heavily on educational data mining, we can pave the way for a more inclusive and effective education system that caters to the unique needs of every student.

Instructional Design and Curriculum Improvement

In the rapidly evolving landscape of education, instructional design and curriculum improvement play pivotal roles in achieving effective and personalized learning experiences for every individual. This subchapter explores the intersection of these two essential components and highlights the significant impact of data mining in advancing education.

Instructional design is the systematic approach to creating and delivering educational experiences that promote effective learning. It involves analyzing learners' needs, understanding the subject matter, and employing appropriate strategies and technologies to design engaging and impactful learning experiences. By leveraging data mining techniques, instructional designers can gain valuable insights into learners' behaviors, preferences, and learning styles. This data-driven approach enables them to tailor instructional materials and activities to match individual learners' needs, resulting in improved learning outcomes.

Curriculum improvement focuses on enhancing the content, structure, and delivery of educational programs. It involves continuous evaluation and refinement of the curriculum to ensure alignment with educational goals and the evolving needs of learners. Data mining plays a crucial role in this process by providing educators and curriculum developers with evidence-based insights into the effectiveness of instructional approaches and content. By analyzing large datasets, educators can identify patterns, trends, and areas for improvement, enabling them to make data-informed decisions that enhance the curriculum and optimize learning experiences for every student.

Data mining, as a powerful tool in educational research, helps uncover hidden patterns, correlations, and trends in vast amounts of educational data. By analyzing student performance, engagement, and other relevant data, educators can identify factors that positively or negatively impact learning outcomes. This knowledge empowers them to make data-driven decisions regarding instructional strategies, content selection, and curriculum modifications. Consequently, data mining enables educators to personalize learning experiences, address learning gaps, and provide targeted interventions, ultimately improving student achievement and engagement.

For individuals interested in data mining and its application in education, this subchapter provides valuable insights into how data mining techniques can inform instructional design and curriculum improvement. By embracing data-driven approaches, educators and curriculum developers can harness the power of educational data mining to create more personalized, effective, and engaging learning experiences for every individual.

In conclusion, instructional design and curriculum improvement are integral to the future of education. Data mining, as a key component of these processes, enables educators to gain insights into learners' needs, refine instructional practices, and tailor curricula to optimize learning outcomes. By embracing data-driven approaches, every individual in the field of education can contribute to the advancement of personalized and effective learning experiences for all.

Educational Policy and Decision Making

In today's rapidly evolving world, education plays a pivotal role in shaping the future of individuals and societies. As we move towards a more data-driven era, the field of educational data mining has emerged as a powerful tool to inform policy-making and decision-making processes within the education system. This subchapter delves into the crucial intersection of education, data mining, and how it can revolutionize educational policies and decisions for the betterment of every individual.

Educational policy-making is a complex and multifaceted process that involves the development, implementation, and evaluation of policies that govern various aspects of education. Traditionally, these policies have been created based on anecdotal evidence, expert opinions, and limited data. However, with the advent of data mining techniques, policymakers now have access to vast amounts of educational data that can provide valuable insights into student learning, performance, and success factors.

Data mining in education involves the extraction of knowledge and patterns from large datasets to uncover hidden relationships and trends. By analyzing this data, policymakers can gain a deeper understanding of the effectiveness of existing policies, identify areas for improvement, and make informed decisions that can positively impact the learning outcomes of every individual.

One key benefit of educational data mining is its ability to identify and address educational inequalities. By analyzing data on student performance, demographics, and socio-economic factors, policymakers can pinpoint disparities in educational opportunities and outcomes. This information can then be used to develop targeted interventions and policies that aim to bridge the gaps and ensure equal access to quality education for all.

Furthermore, data mining can assist in designing personalized learning experiences tailored to individual students' needs. By analyzing student data, such as learning preferences, strengths, and weaknesses, policymakers can create adaptive learning environments that enhance student engagement and achievement. This approach recognizes the unique characteristics and learning styles of every individual, fostering a more inclusive and effective education system.

However, it is crucial to address the ethical and privacy concerns associated with educational data mining. Safeguarding sensitive student information and ensuring data is used responsibly should be paramount. Striking the right balance between data-driven decision-making and protecting privacy is essential to maintain public trust and ensure the long-term success of educational policies.

In conclusion, the integration of data mining in educational policy-making and decision-making processes holds immense potential to transform education for everyone. By leveraging the power of data, policymakers can make evidence-based decisions, address inequalities, and create a more personalized and inclusive learning environment. It is imperative to embrace this data-driven future while upholding ethical considerations to shape a brighter future for education.

Chapter 6: Case Studies in Educational Data Mining

Case Study 1: Improving Student Performance with Data Mining

In today's rapidly evolving world, data mining has emerged as a powerful tool in various industries. One particular field where data mining has shown immense potential is education. By harnessing the power of this technology, educators can now gain valuable insights into student performance, enabling them to make data-driven decisions and drive positive outcomes for every individual.

This case study delves into the transformative impact of data mining on student performance, highlighting its significance in shaping the future of learning. Addressed to everyone, especially those interested in data mining, this subchapter sheds light on the potential benefits and challenges associated with this approach.

Data mining involves the extraction and analysis of large amounts of data to identify patterns, trends, and correlations. In the context of education, it enables educators to collect and analyze vast amounts of student data, including academic records, test scores, attendance, and even social interactions, to gain a holistic understanding of each student's learning journey.

By utilizing data mining techniques, educators can identify patterns that may have otherwise gone unnoticed. They can identify specific areas where students are struggling, pinpoint effective teaching methods, and tailor instruction to meet individual needs. Data mining also allows for early identification of at-risk students, enabling timely intervention and support, ultimately improving overall student performance.

One example of data mining's impact on student performance comes from a high school in a diverse urban setting. By analyzing student data, the school discovered that a significant number of students were failing to complete assignments and were experiencing a decline in grades. Through data mining, the school identified that these students were consistently absent on Mondays, leading to a lack of continuity in their learning.

Armed with this information, the school implemented targeted interventions, such as engaging students in Monday morning activities and providing additional support for missed coursework. As a result, student attendance improved, assignment completion rates increased, and overall grades showed a marked improvement.

However, data mining in education also presents challenges that need to be addressed. Privacy concerns, data security, and ethical use of data are critical considerations when implementing data mining techniques in educational settings. It is imperative to strike the right balance between data collection and protecting students' privacy rights.

In conclusion, data mining has emerged as a powerful tool in improving student performance. By leveraging the insights gained from data analysis, educators can create personalized learning experiences, identify at-risk students, and make data-driven decisions that positively impact student outcomes. However, it is essential to address the challenges associated with data mining to ensure its responsible and ethical implementation in educational settings. The future of learning lies in leveraging data mining effectively to create a more inclusive and personalized educational experience for every individual.

Case Study 2: Enhancing Instructional Effectiveness through Data Analysis

In today's digital age, the collection and analysis of data have become crucial in almost every industry, including education. With the advent of educational data mining, educators now have a powerful tool to enhance instructional effectiveness and improve learning outcomes for every individual. This subchapter delves into a captivating case study that highlights the transformative potential of data analysis in education.

The case study revolves around a diverse group of students at a secondary school who were struggling with their math performance. The school administration decided to employ data mining techniques to identify the underlying causes of these challenges and develop customized interventions to support each student's unique needs.

The first step involved collecting a vast amount of data, including students' test scores, attendance records, and engagement levels in class. By leveraging advanced data mining algorithms, patterns and trends within this data were then identified. The analysis revealed that specific mathematical concepts were consistently problematic for the majority of students, indicating a need for targeted instruction in these areas.

Armed with this information, the educators developed an intervention plan tailored to each student's individual requirements. Utilizing personalized learning platforms and adaptive software, students were provided with interactive resources and activities designed to address their specific learning gaps. Furthermore, regular progress assessments were conducted to monitor their growth and refine the instructional strategies accordingly.

The results were astounding. Over the course of the year, the students' math scores witnessed a remarkable improvement. The analysis of ongoing data also allowed educators to identify students who required additional support and provided timely interventions to ensure their progress was not hindered.

This case study exemplifies the power of data mining in education. By analyzing vast amounts of data, educators can gain valuable insights into students' learning patterns, identify areas of improvement, and develop targeted interventions to foster their academic growth. Moreover, the integration of personalized learning platforms and adaptive software enables educators to create tailored learning experiences that cater to individual needs, ensuring no student is left behind.

In conclusion, educational data mining has the potential to revolutionize the way we approach instruction. By harnessing the power of data analysis, educators can enhance instructional effectiveness, promote personalized learning, and ultimately empower every individual to reach their full potential. As we embrace the future of learning, it is imperative that we recognize the significance of data mining and its role in shaping the educational landscape.

Case Study 3: Identifying Learning Patterns for Personalized Interventions

Introduction:

In this chapter, we delve into a fascinating case study that highlights the power of educational data mining in identifying learning patterns for personalized interventions. As technology continues to revolutionize the education sector, data mining plays a pivotal role in unlocking valuable insights that can transform the way individuals learn. This case study explores the potential of data mining to create personalized interventions, catering to the unique needs and learning styles of every individual.

Understanding Educational Data Mining: Educational data mining is the process of extracting meaningful patterns and trends from large datasets generated by educational systems and institutions. By analyzing this vast amount of data, educators can gain invaluable insights into student behavior, learning patterns, and performance. This information enables them to develop personalized interventions that enhance the learning experience and improve educational outcomes.

The Power of Personalized Interventions: Traditionally, education has followed a one-size-fits-all approach, assuming that all students learn in the same way. However, this approach fails to consider the individual differences and unique learning needs of each student. By harnessing the power of data mining, educators can now identify specific learning patterns and tailor interventions accordingly.

Case Study: Identifying Learning Patterns: In this case study, we explore how a renowned educational institution utilized data mining techniques to identify learning patterns among its diverse student population. By collecting and analyzing data from various sources such as online platforms, learning management systems, and assessments, the institution was able to identify distinct patterns of engagement, knowledge acquisition, and skill development.

Personalized Interventions for Optimal Learning: Using the insights gained from data mining, the institution was able to develop personalized interventions for each student. These interventions ranged from adaptive learning platforms that adjusted the difficulty level of coursework based on individual progress to targeted feedback and additional resources tailored to address specific learning gaps. By tailoring interventions to individual needs, the institution witnessed improved engagement, motivation, and overall academic performance among its students.

Conclusion:
Data mining is revolutionizing the field of education by enabling personalized interventions that cater to the unique learning patterns of every individual. This case study demonstrates the potential of educational data mining in transforming traditional educational approaches into personalized, adaptive systems that optimize learning outcomes. As data mining continues to evolve, we can expect even more innovative interventions that revolutionize the future of learning for everyone.

Chapter 7: Future Trends and Challenges in Educational Data Mining

Emerging Technologies in Educational Data Mining

In recent years, the field of educational data mining has witnessed significant advancements due to the integration of emerging technologies. These technologies have revolutionized the way we collect, analyze, and utilize data in the realm of education. In this subchapter, we will explore some of the most promising emerging technologies in educational data mining and their potential to shape the future of learning.

One of the key technologies that have gained traction in educational data mining is machine learning. Machine learning algorithms have the ability to analyze large datasets and identify patterns, trends, and correlations that may not be apparent to human observers. By applying machine learning techniques to educational data, educators can gain valuable insights into student performance, preferences, and learning styles. This information can then be used to personalize instruction, design targeted interventions, and optimize learning resources.

Another emerging technology that holds great promise in educational data mining is natural language processing (NLP). NLP allows computers to understand and interpret human language, enabling the analysis of large volumes of textual data such as essays, forum discussions, and feedback forms. By employing NLP techniques, educators can extract meaningful information from textual data, such as identifying common misconceptions, assessing the quality of student writing, or detecting sentiment and emotions. This information can provide valuable feedback to both students and teachers, aiding in the improvement of learning outcomes.

Virtual reality (VR) and augmented reality (AR) are also emerging technologies that have the potential to transform the educational landscape. Through VR and AR, students can engage in immersive and interactive learning experiences, allowing them to visualize complex concepts, explore virtual environments, and practice real-world skills in a safe and controlled setting. By integrating these technologies with data mining techniques, educators can capture and analyze student interactions within these virtual environments, providing insights into their learning processes and identifying areas for improvement.

Lastly, the field of educational data mining is also witnessing the rise of blockchain technology. Blockchain, best known for its association with cryptocurrencies, offers the potential to securely store and verify educational credentials, certifications, and achievements. By leveraging blockchain, educational institutions can ensure the integrity and authenticity of student records, making them tamper-proof and easily shareable. This technology has the potential to revolutionize the way educational data is stored, managed, and shared, empowering individuals to have greater control over their educational achievements.

In conclusion, emerging technologies in educational data mining are bringing about transformative changes in the way we understand and facilitate learning. Machine learning, natural language processing, virtual reality, augmented reality, and blockchain are just a few examples of technologies that hold immense potential in enhancing educational practices. As we move forward, it is crucial for educators, researchers, and policymakers to embrace these technologies and leverage their capabilities to create a future of learning that is personalized, engaging, and inclusive for every individual.

Ethical Considerations in the Future of Educational Data Mining

In today's data-driven world, the field of educational data mining (EDM) has emerged as a powerful tool to enhance learning outcomes for individuals of all ages. As we delve deeper into the future of educational data mining, it is crucial to address the ethical considerations that arise with the use of such technology. This subchapter aims to shed light on the potential ethical challenges and provide guidance on how to navigate them.

First and foremost, one of the primary ethical considerations in EDM is data privacy. Educational institutions gather vast amounts of data from students, including their academic performance, personal information, and even their online behavior. It is imperative to ensure that this data is collected and stored securely, with the consent of the individuals involved. Strict protocols for data anonymization and encryption must be in place to protect the privacy of students and prevent any potential misuse.

Transparency is another critical ethical consideration. It is essential to be open and transparent about the collection and use of educational data. Individuals should have a clear understanding of what data is being collected, how it will be used, and who will have access to it. Providing individuals with control over their data, such as the ability to opt-out or delete their information, is crucial to uphold ethical standards.

Fairness and equity are also paramount in the future of educational data mining. The algorithms and models used in EDM must be designed with sensitivity to potential bias, ensuring that they do not perpetuate existing inequalities or discriminate against certain individuals or groups. Constant monitoring and auditing of the algorithms are necessary to identify and rectify any biases that may emerge.

Furthermore, the ethical use of educational data mining involves the responsible application of the insights derived from the data. The data collected should be utilized to provide personalized and inclusive educational experiences, rather than solely for profit-driven purposes. The focus should be on empowering individuals and improving learning outcomes, rather than exploiting their data for commercial gain.

To address these ethical considerations, collaboration between educational institutions, data scientists, policymakers, and other stakeholders is crucial. Establishing clear guidelines and ethical frameworks is necessary to ensure that the future of educational data mining is approached with responsibility and integrity.

In conclusion, while the future of educational data mining holds immense potential for enhancing learning outcomes, it is vital to address the ethical considerations associated with its use. Data privacy, transparency, fairness, equity, and responsible application of insights should be at the forefront of every decision made in this field. By upholding these ethical standards, we can ensure that educational data mining benefits every individual and contributes to a more equitable and inclusive education system.

Challenges and Limitations in Implementing Educational Data Mining

As the field of data mining continues to expand, its application in the educational domain has gained significant attention. Educational data mining (EDM) aims to uncover meaningful patterns and insights from large datasets collected in educational settings. By analyzing this data, educators and researchers can gain valuable insights into student learning behaviors, academic performance, and instructional effectiveness. However, like any emerging field, EDM faces its own set of challenges and limitations that need to be addressed for its successful implementation.

One of the primary challenges in implementing EDM is the ethical and privacy concerns associated with collecting and analyzing student data. With the increasing use of technology in classrooms, vast amounts of data are being generated on a daily basis. This data includes students' personal and sensitive information, and there is a need to ensure that it is protected and used appropriately. Privacy regulations and policies must be established to safeguard student data and maintain the trust of all stakeholders involved.

Another challenge is the lack of standardized data collection methods and tools. Educational institutions often collect data in different formats and using various systems, making it difficult to compare and integrate datasets. The lack of standardization hampers the ability to conduct large-scale studies and limits the generalizability of research findings. Efforts should be made to develop standardized data collection protocols and tools that can be widely adopted across educational settings.

Furthermore, the complexity and multidimensionality of educational datasets pose a challenge in extracting meaningful insights. Educational data is often characterized by its heterogeneity, including variables such as student demographics, academic performance, learning activities, and social interactions. Analyzing such complex datasets requires advanced statistical techniques and data mining algorithms. Training educators and researchers in these methods and providing them with the necessary tools and resources is crucial for leveraging the full potential of EDM.

Additionally, the implementation of EDM faces limitations in terms of scalability and resource requirements. Analyzing large datasets requires significant computational power and storage capacity. Educational institutions may face challenges in acquiring and maintaining the necessary infrastructure. Moreover, there may be a shortage of skilled data scientists and analysts who can effectively handle and interpret educational data. Investment in resources and training is essential to overcome these limitations and ensure the successful implementation of EDM.

In conclusion, while educational data mining holds immense potential for improving teaching and learning, it is not without its challenges and limitations. Addressing ethical concerns, standardizing data collection methods, developing advanced analytical techniques, and investing in resources are crucial steps towards realizing the full benefits of EDM. By overcoming these challenges, educators, researchers, and policymakers can harness the power of data mining to drive evidence-based decision-making, enhance educational outcomes, and shape the future of learning for every individual.

Chapter 8: Conclusion

Recap of Key Concepts

In this subchapter, we will recap the key concepts discussed throughout the book "The Future of Learning: Educational Data Mining for Every Individual." Whether you are a data mining expert or new to the field, this recap will serve as a useful reference to reinforce your understanding of the book's main ideas.

Educational Data Mining (EDM) serves as the foundation for the future of learning. It involves the application of data mining techniques to educational data, providing valuable insights into students' learning patterns, performance, and preferences. By analyzing this data, educators can personalize instruction, identify at-risk students, and improve overall learning outcomes.

One crucial concept in EDM is the collection and analysis of Big Data. Educational institutions generate vast amounts of data, including student demographics, grades, attendance records, and even online learning interactions. By harnessing the power of Big Data, institutions can gain a comprehensive understanding of each student's unique learning journey, allowing for targeted interventions and support.

Another key concept is the use of predictive modeling to anticipate student outcomes. By analyzing historical data, algorithms can identify patterns and make predictions about a student's future performance. This enables educators to intervene early and provide personalized support to ensure student success.

Data visualization plays a vital role in EDM. It enables educators to present complex data in an easily understandable format. By visualizing learning analytics, educators can identify trends, patterns, and outliers, aiding decision-making processes and improving educational practices.

Ethical considerations are crucial in the field of data mining. Privacy concerns and data protection must be addressed to ensure that students' information is handled with utmost care. This includes obtaining informed consent, anonymizing data, and implementing secure data storage and transmission protocols.

Lastly, the integration of EDM into Learning Management Systems (LMS) is essential for its widespread adoption. LMS platforms can collect and analyze data in real-time, providing immediate feedback to both students and educators. This integration enhances the learning experience, promotes self-regulation, and enables personalized learning pathways.

In conclusion, "The Future of Learning: Educational Data Mining for Every Individual" introduces the field of EDM and its potential to revolutionize education. By leveraging data mining techniques, institutions can gain valuable insights into student learning, personalize instruction, and improve educational outcomes. However, it is crucial to address ethical considerations and integrate EDM into existing educational systems to ensure its effective implementation.

Implications and Potential of Educational Data Mining

In the rapidly evolving world of education, data mining has emerged as a powerful tool with immense potential to transform the way we learn and teach. Educational data mining is the process of extracting valuable insights and patterns from large sets of educational data, ranging from student performance records to online learning interactions. This subchapter explores the implications and potential of educational data mining, shedding light on the significant impact it can have on every individual, regardless of their background or field of interest.

One of the key implications of educational data mining is its ability to personalize the learning experience. By analyzing vast amounts of data, educators can gain a deep understanding of each student's strengths, weaknesses, and learning styles. This knowledge can be leveraged to tailor instructional strategies and provide individualized support, ensuring that every student has the opportunity to thrive. Through personalized learning, educational data mining has the potential to bridge the achievement gap, empower struggling learners, and unleash the full potential of every individual.

Another implication of educational data mining lies in its role in shaping educational policies and reforms. By analyzing data collected from diverse sources, policymakers can gain valuable insights into the effectiveness of different teaching methods, curriculum designs, and assessment strategies. This evidence-based approach to decision-making can lead to more informed policy choices, ultimately improving the quality of education at a systemic level. Educational data mining can also help identify and address systemic inequalities, highlighting areas where interventions are most needed to ensure equitable learning opportunities for all.

Furthermore, educational data mining holds immense potential for innovation in educational technology. By analyzing patterns in student behavior and engagement, developers can create adaptive learning platforms that dynamically adjust to individual needs. These platforms can provide real-time feedback, suggest personalized learning resources, and foster self-directed learning. Moreover, educational data mining can contribute to the development of intelligent tutoring systems, virtual reality simulations, and other cutting-edge technologies that enhance the learning experience.

In conclusion, educational data mining has transformative implications and vast potential for every individual and the field of data mining. By personalizing learning, informing policy decisions, and driving technological innovation, educational data mining can revolutionize education, making it more inclusive, effective, and engaging for learners of all backgrounds. As we embark on the future of learning, embracing the power of educational data mining is crucial to unlocking the full potential of every learner and shaping a brighter future for education as a whole.

The Future of Learning: Advancements in Educational Data Mining

In this digital age, the field of education is undergoing a significant transformation. With the rapid advancements in technology, the way we learn and teach has evolved tremendously. One of the most promising areas of innovation is the use of educational data mining. This subchapter explores the future of learning and how data mining can revolutionize education for everyone.

Data mining refers to the process of extracting valuable insights and patterns from large datasets. In the context of education, it involves analyzing vast amounts of student data to uncover meaningful information. By harnessing the power of data mining, educators can gain valuable insights into students' learning patterns, preferences, and needs.

The future of learning lies in personalized education. Every individual has unique strengths, weaknesses, and learning styles. Educational data mining enables us to create customized learning experiences tailored to each student's needs. By analyzing data on student performance, engagement, and interactions, educators can identify areas where students may be struggling and provide targeted interventions.

Furthermore, data mining can help identify effective teaching strategies and tools. By analyzing data on instructional methods, curriculum design, and assessments, educators can gain insights into what works best for different students. This data-driven approach to teaching can lead to more effective and efficient educational practices, ultimately improving learning outcomes for all.

Another exciting advancement in educational data mining is the use of predictive analytics. By analyzing historical data, educators can predict future performance and identify students who may be at risk of falling behind. This early identification allows for timely interventions and targeted support, ensuring that no student is left behind.

However, it is essential to address the ethical implications of educational data mining. Privacy concerns and data security should be a top priority. Safeguarding student data and ensuring its responsible use are paramount. Transparency and informed consent should be the guiding principles in implementing educational data mining practices.

In conclusion, the future of learning is bright with the advancements in educational data mining. By leveraging the power of data, educators can create personalized learning experiences, identify effective teaching strategies, and predict future performance. However, it is crucial to approach data mining ethically, with privacy and security at the forefront. With responsible implementation, educational data mining has the potential to transform education, benefiting students of all backgrounds and abilities.